CBIP '75

CHILDREN'S BOOK COUNCIL EXHIBIT
MISSOURI STATE LIBRARY

C4A/BSC 10-74

THE SPIDER WEB

THE SPIDER WEB

by Julie Brinckloe

Doubleday & Company, Inc., Garden City, New York

For my mother and father

ISBN: 0-385-04829-7 Trade
0-385-02821-0 Prebound
Library of Congress Catalog Number 73-20695
Copyright © 1974 by Julie Brinckloe / All Rights Reserved
Printed in the United States of America / First Edition

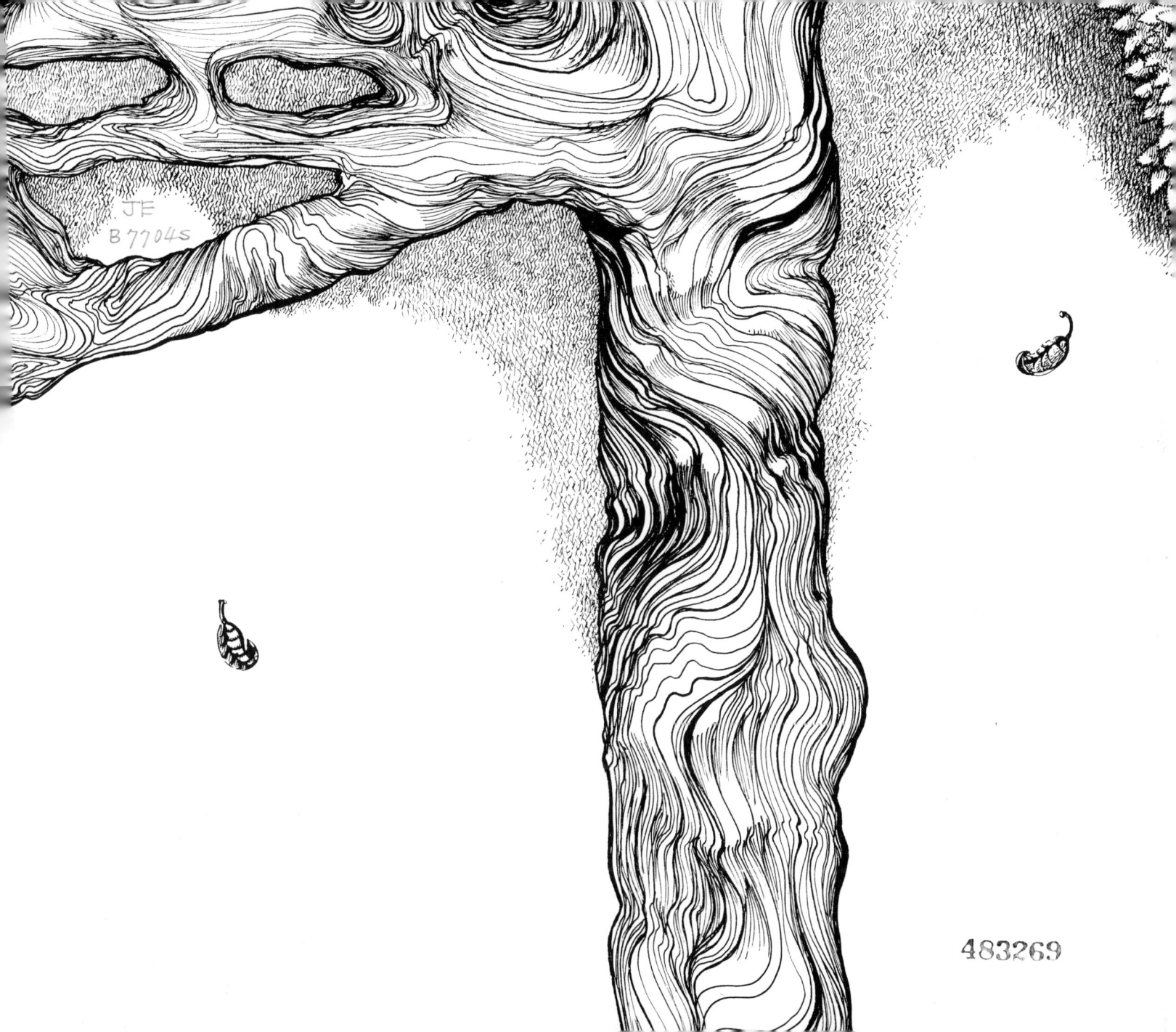

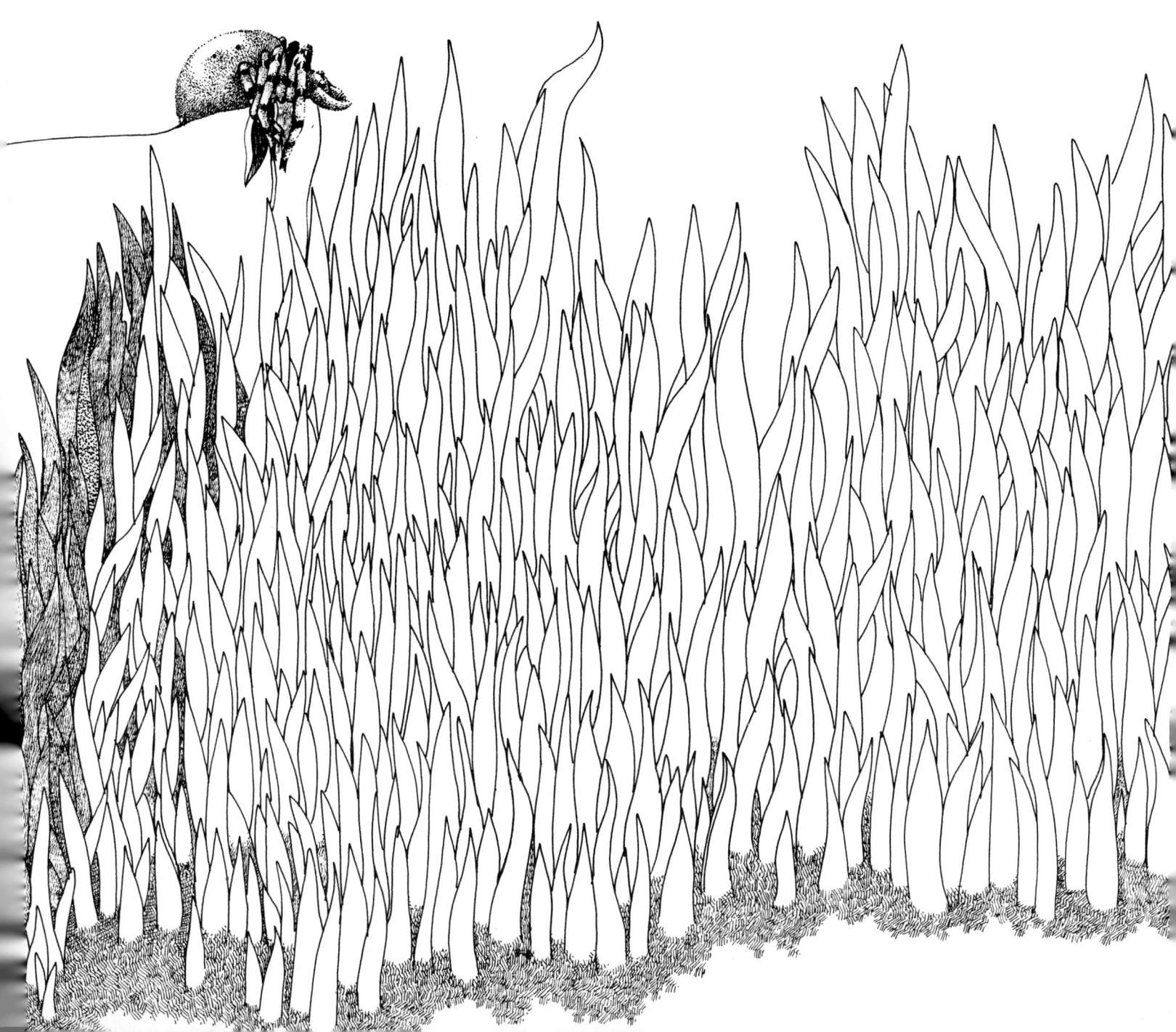

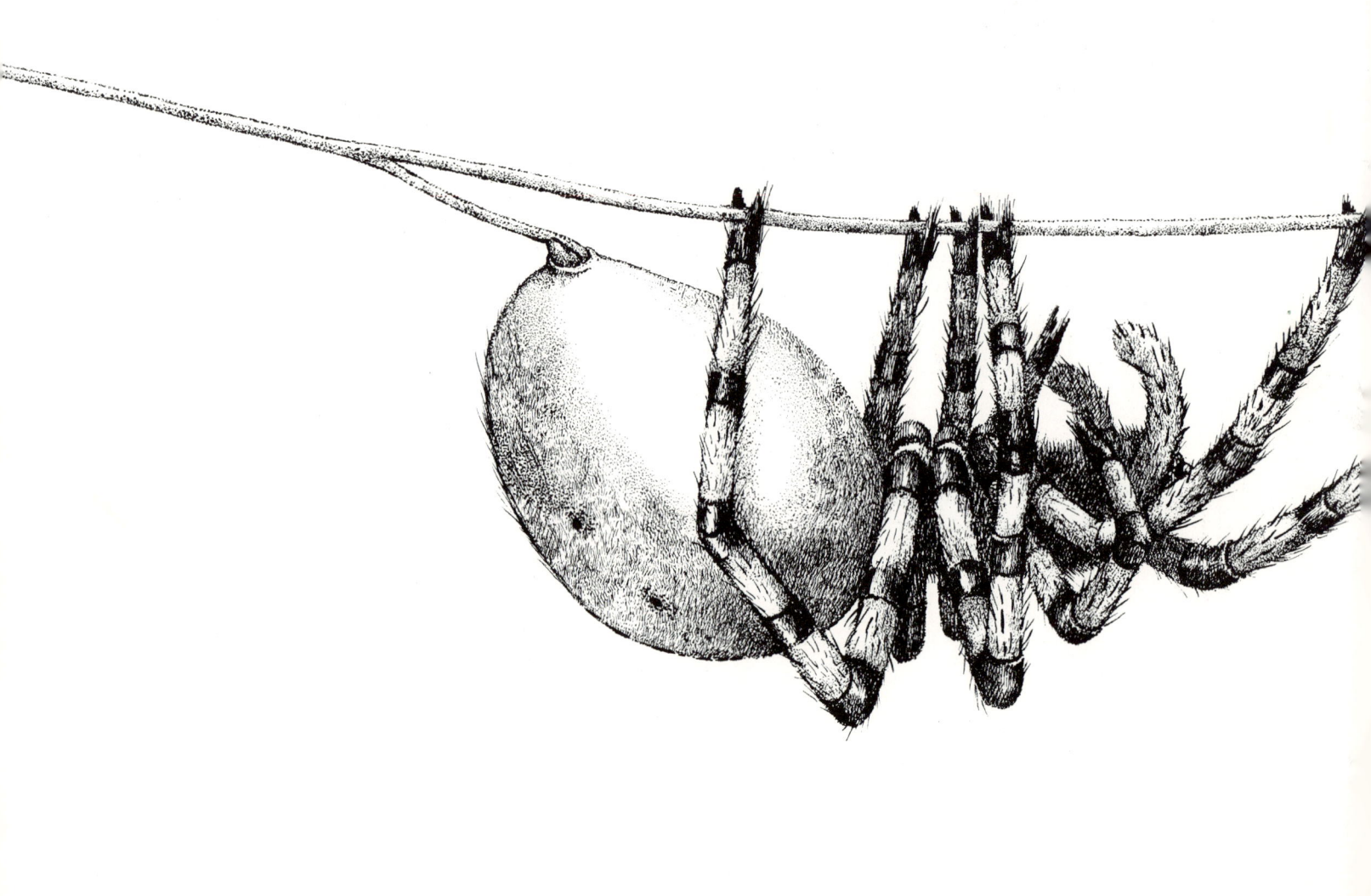

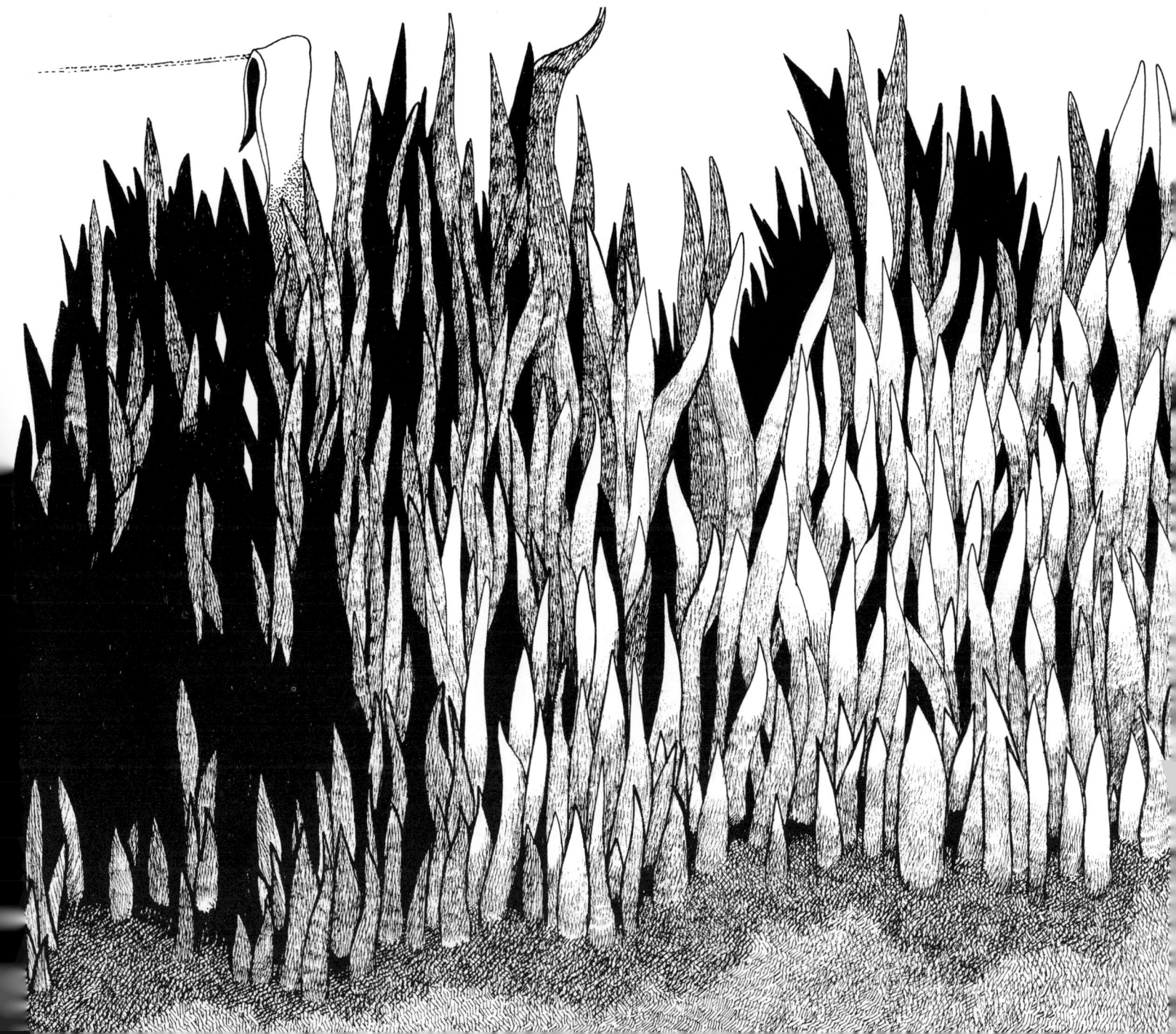

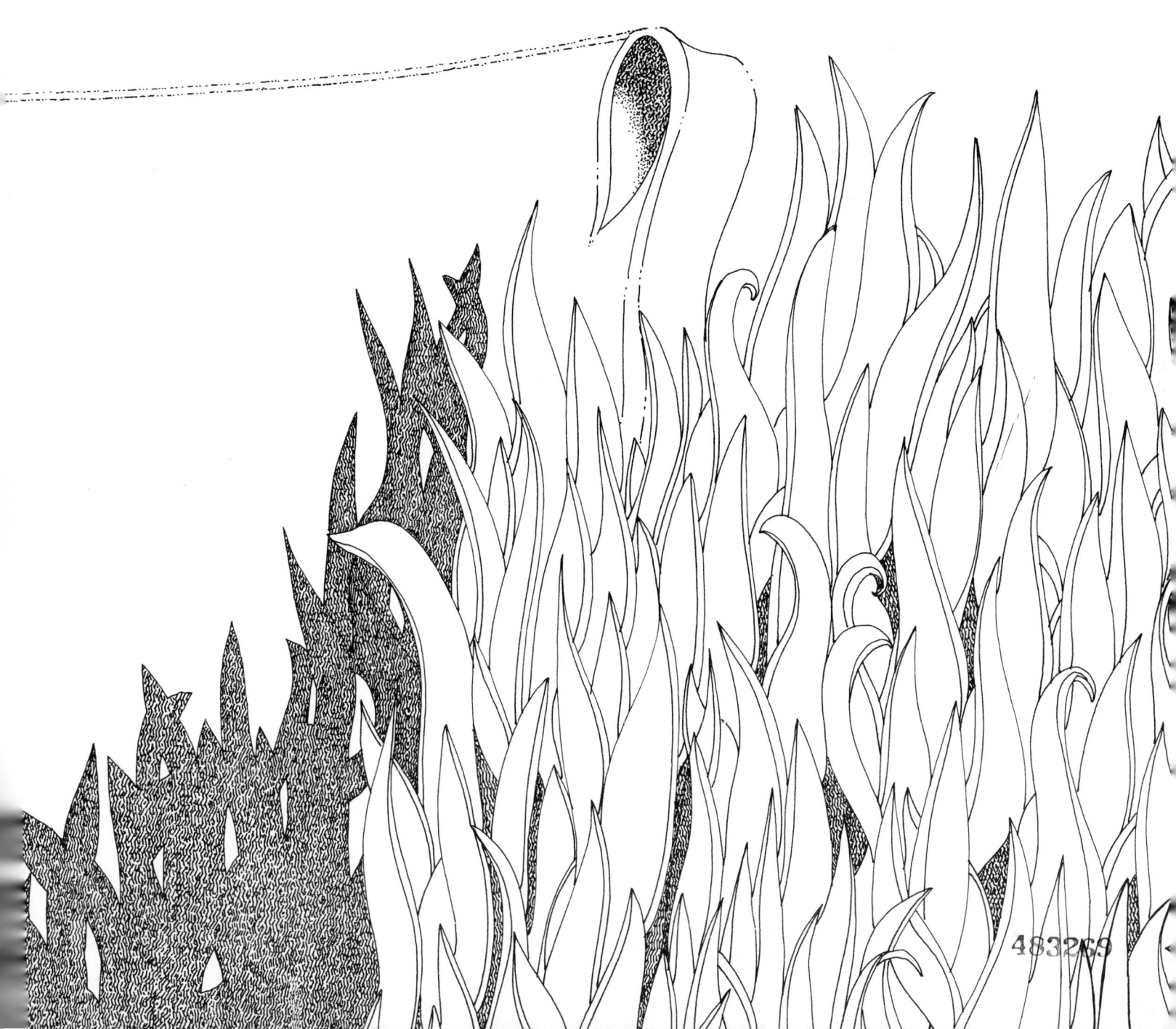

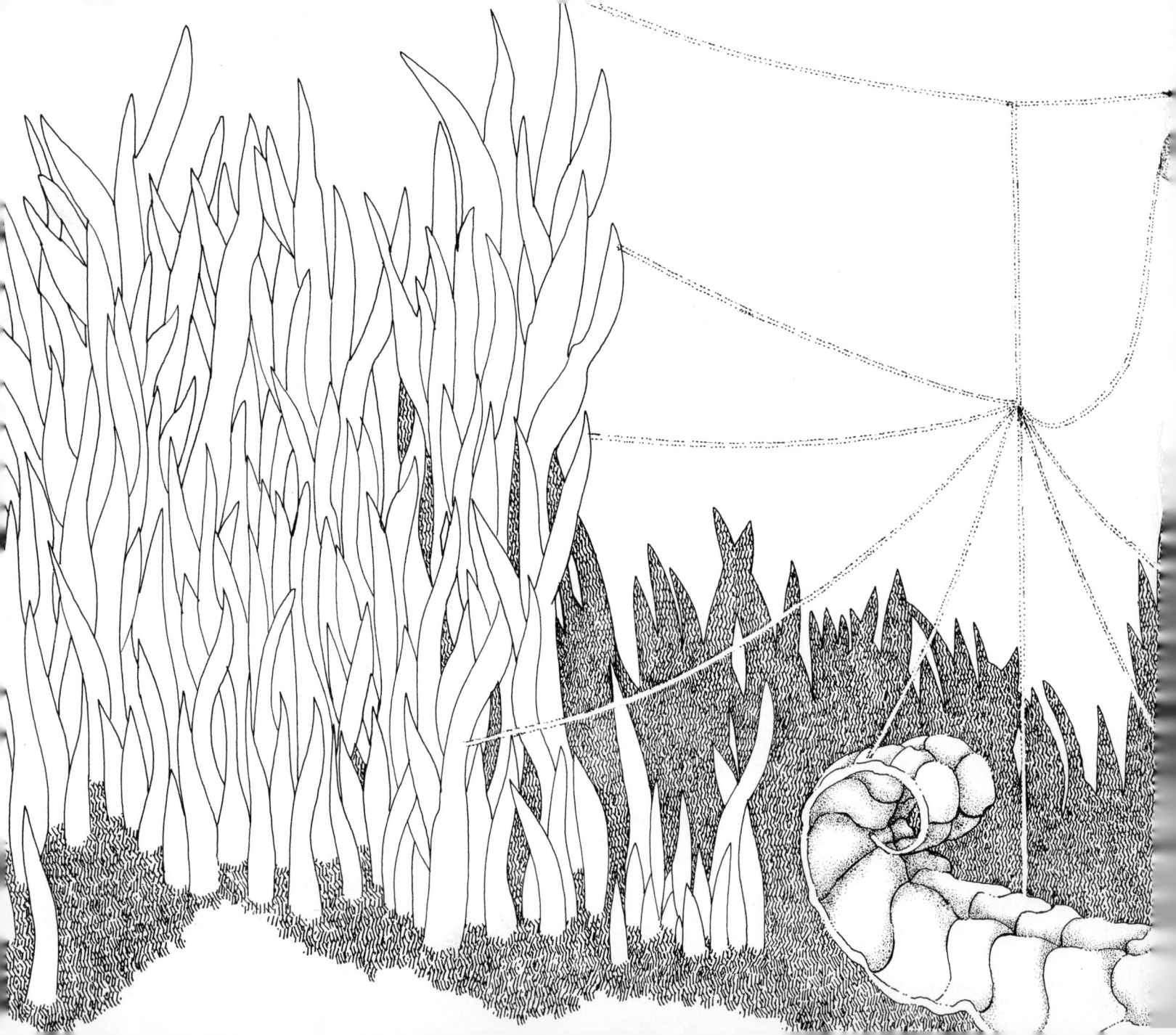

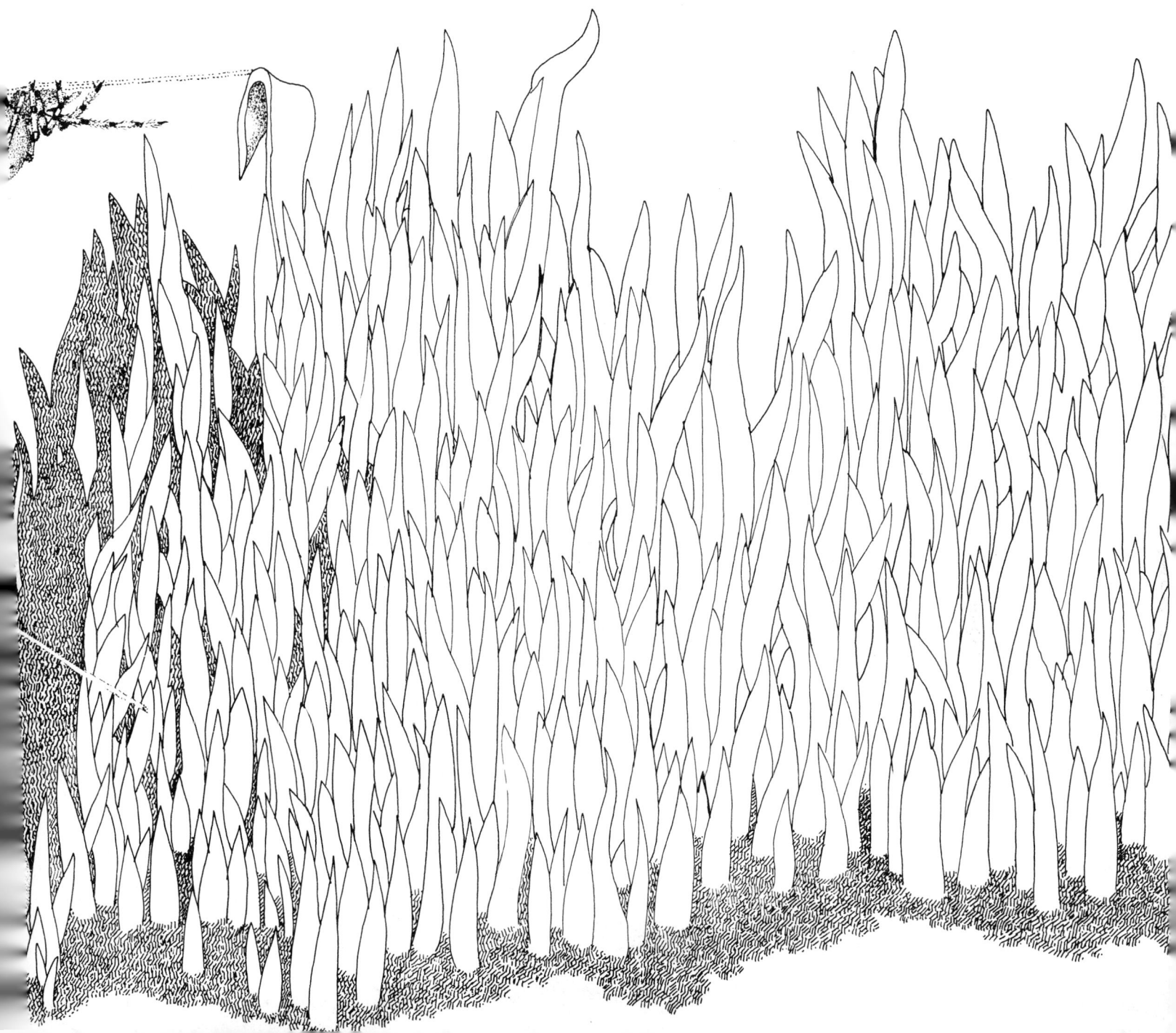

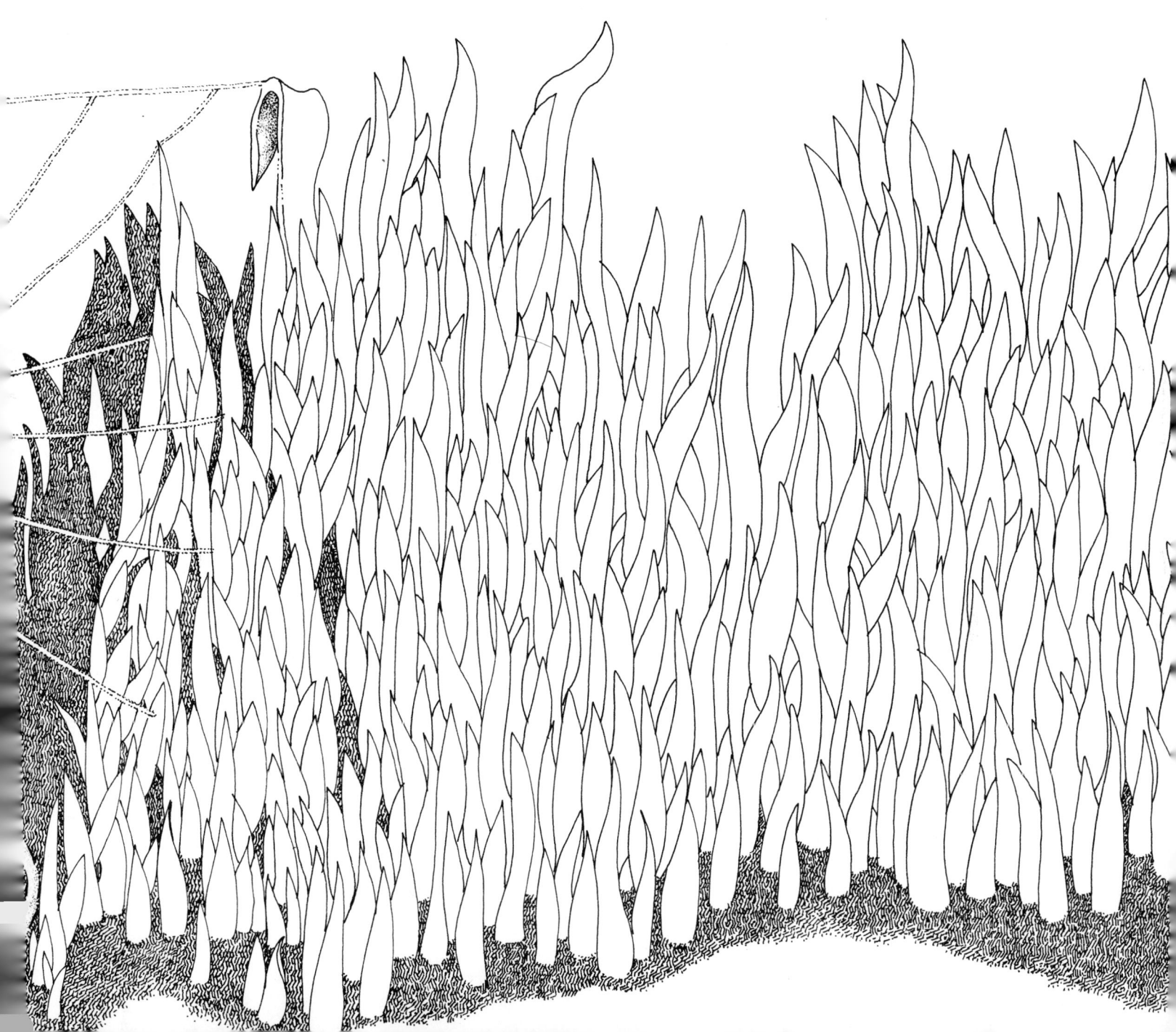

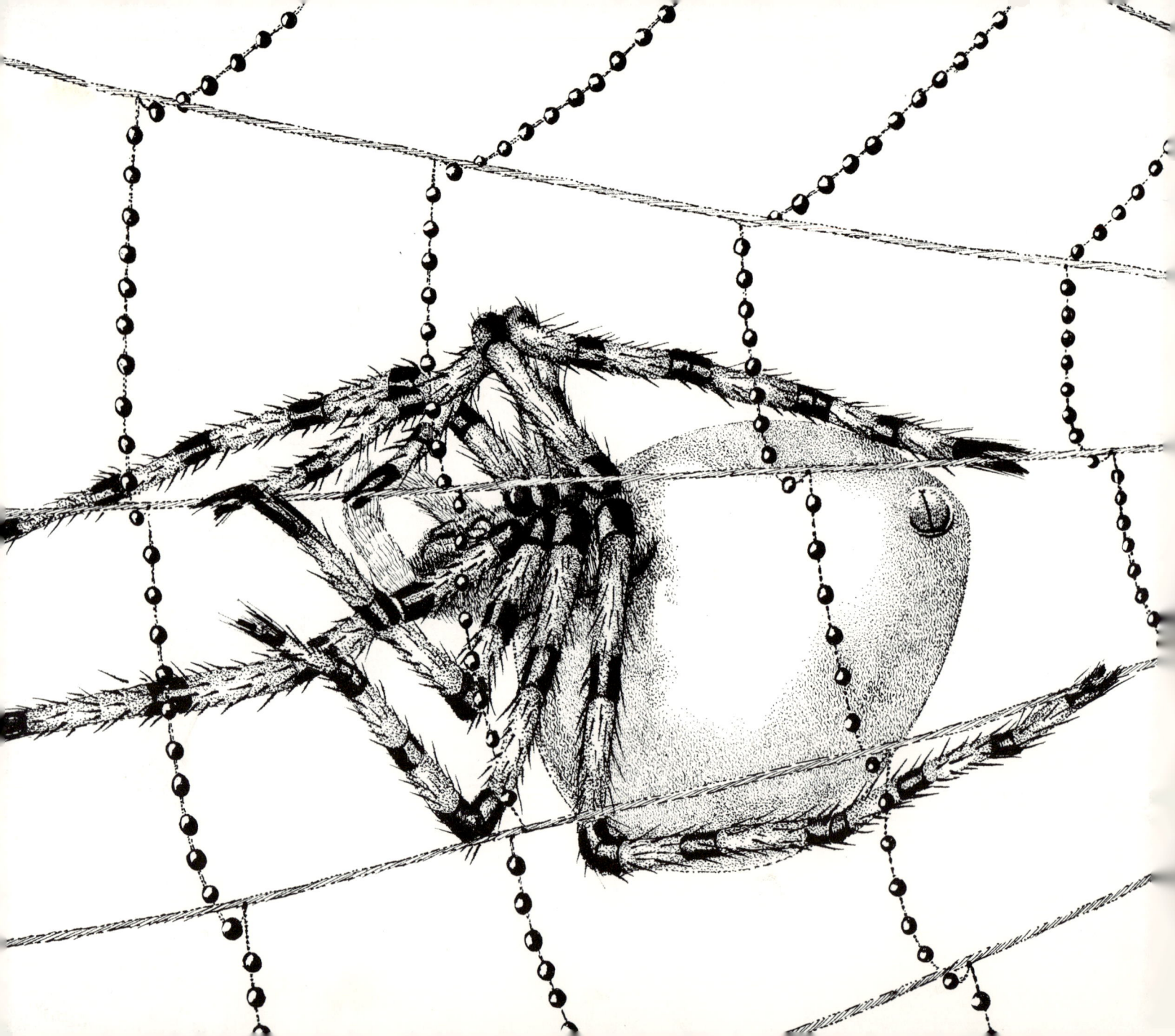

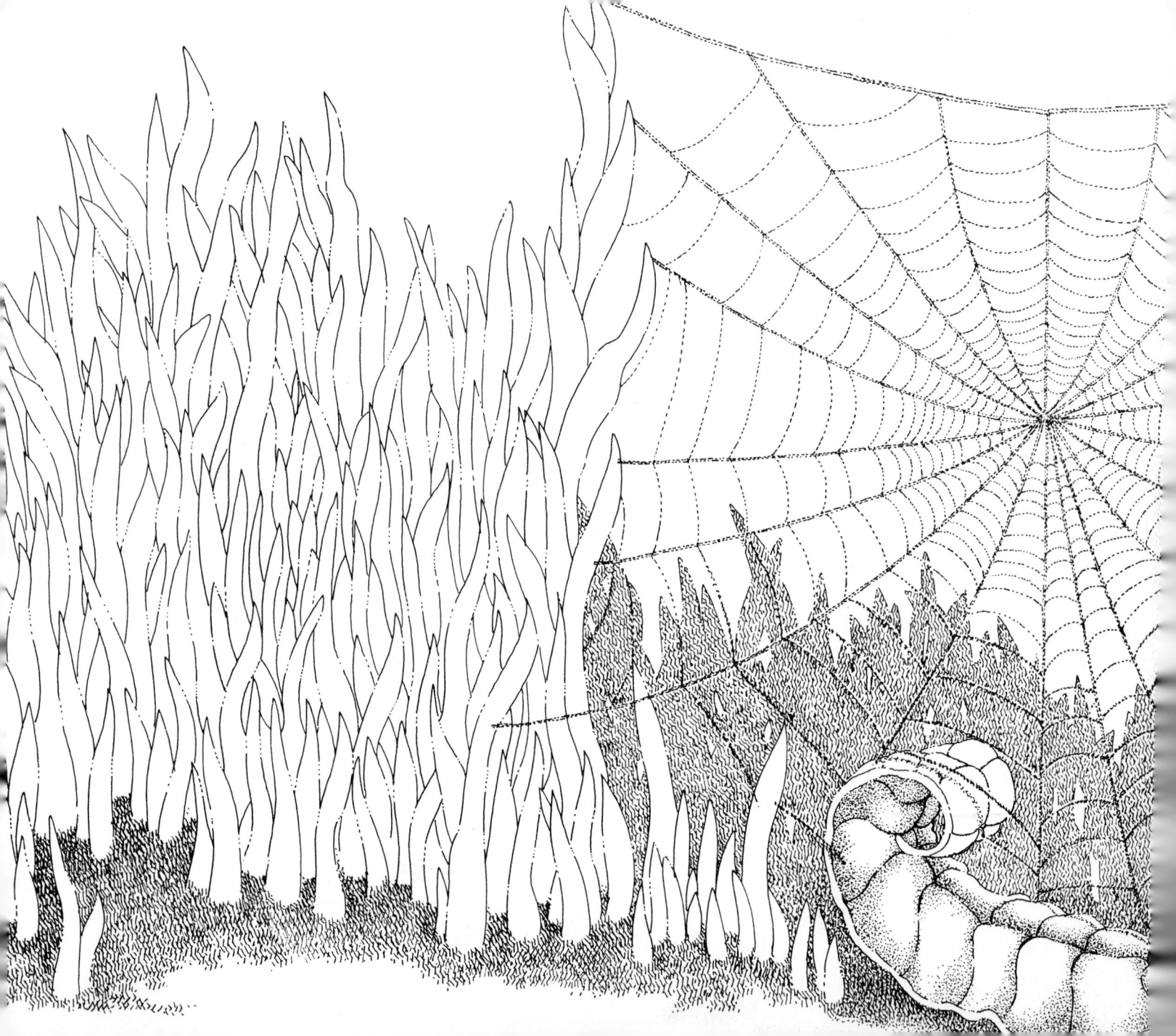

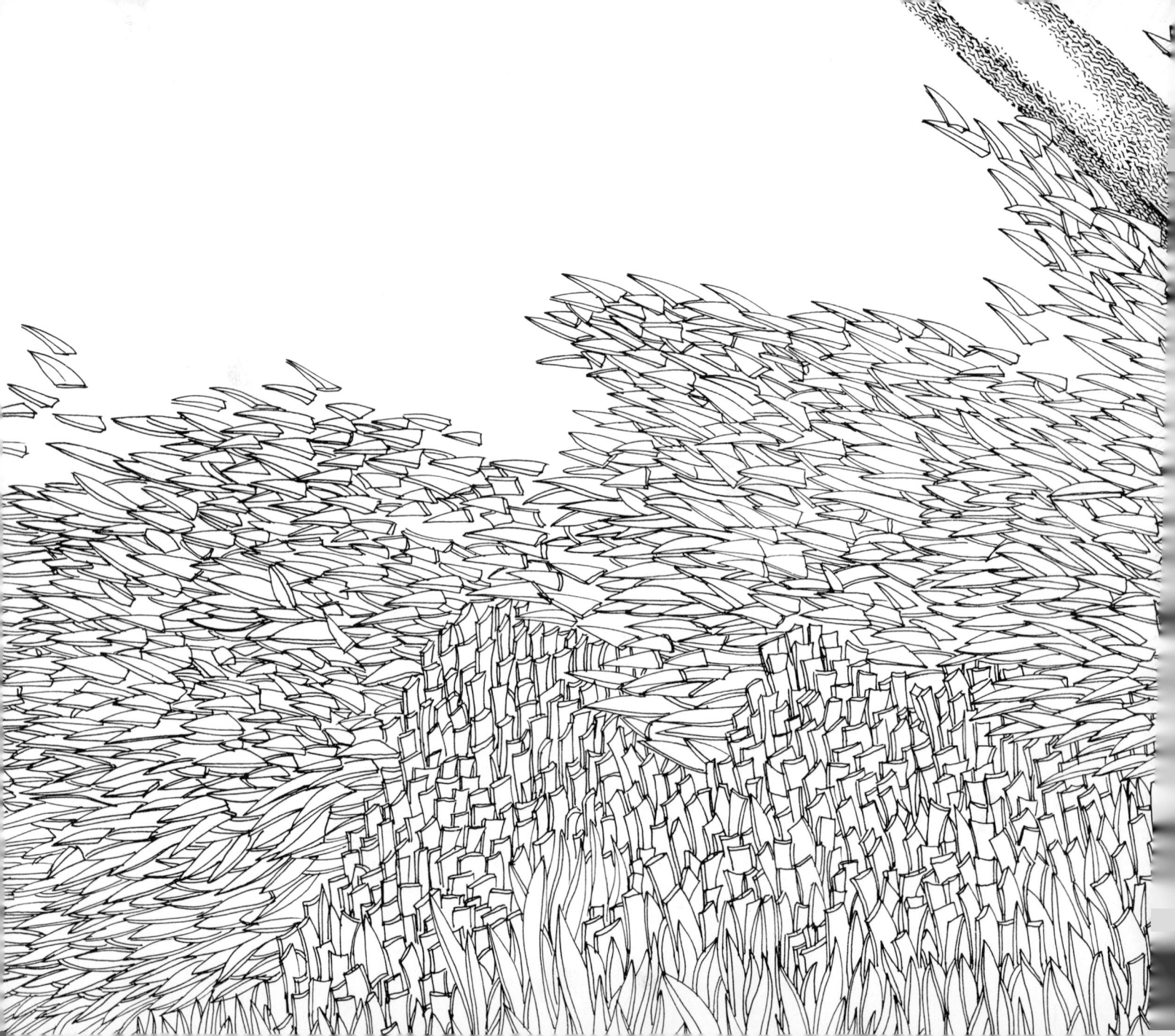

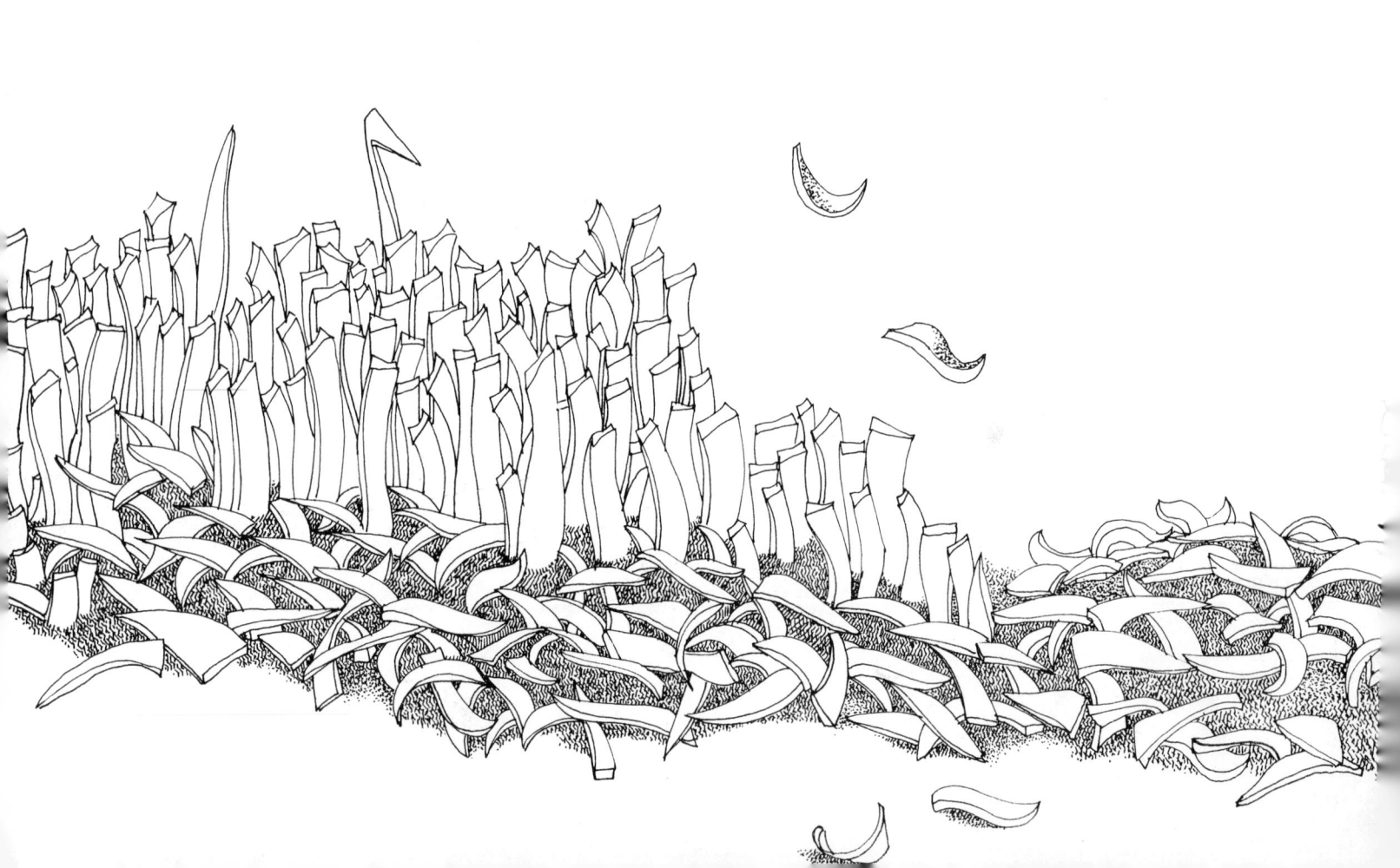

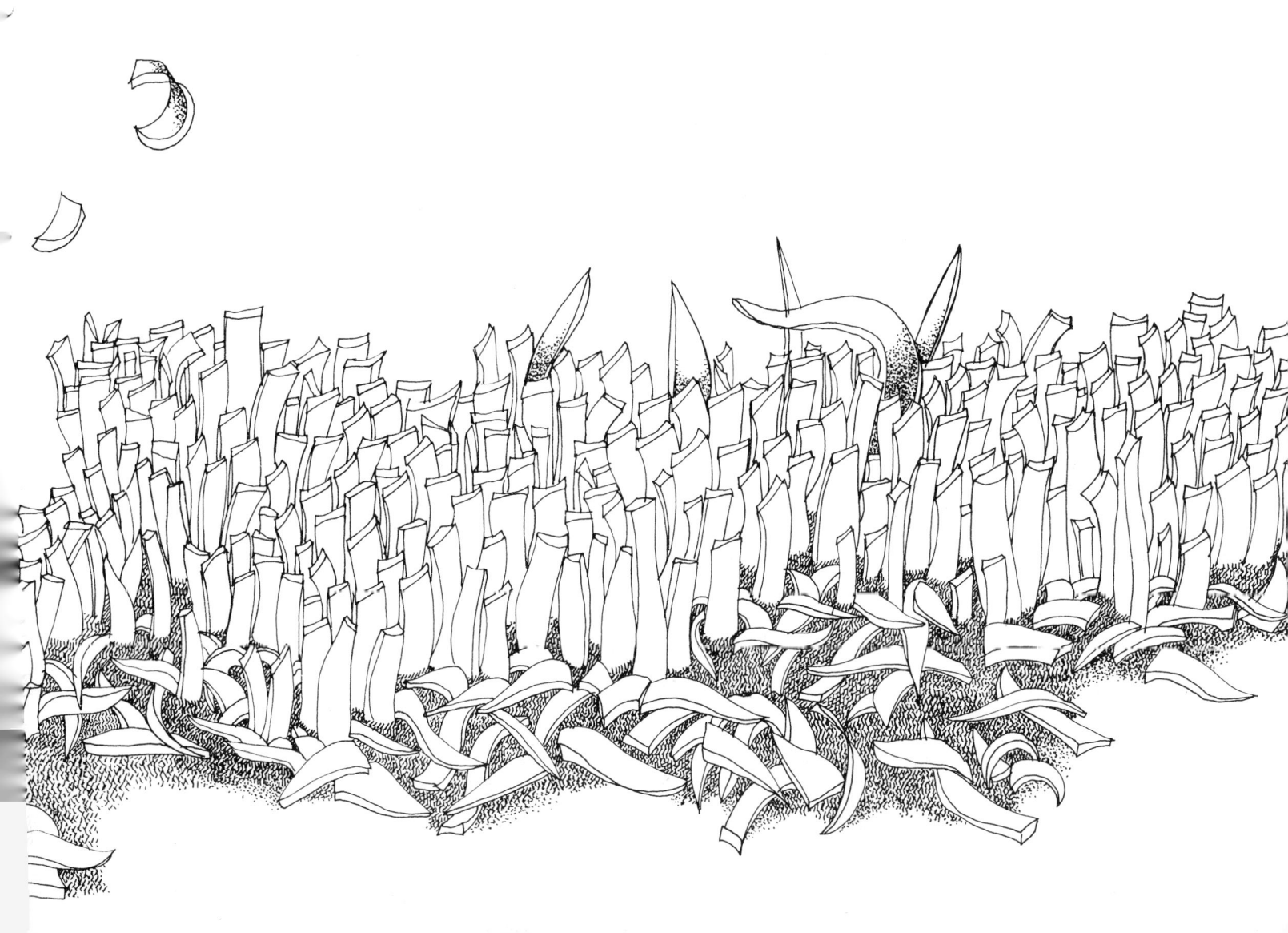

JULIE BRINCKLOE, a native of Mare Island, California, studied art at Sweet Briar College and Carnegie-Mellon University. Although she has illustrated stories and poems by such authors as Art Buchwald (*The Bollo Caper*), Herbert Gold (*The Young Prince and the Magic Cone*) and Theodore Roethke (*Dirty Dinky and Other Creatures: Poems for Children by Theodore Roethke*), *The Spider Web* is her first book to be published. She lives in New York City where she devotes herself full-time to writing and drawing.